Granite Erratics

Also by Brian Bartlett

Cattail Week
Planet Harbor
Underwater Carpentry

Granite Erratics

Brian Bartlett

Ekstasis Editions

Canadian Cataloguing in Publication Data

Bartlett, Brian
Granite erratics

Poems
ISBN 1-896860-15-X

I. Title.
PS8553.A773G73 1997 C811'.54 C97-910235-9
PR9199.3.B374G734 1997

Cover Art: Nelly B. Gray

Acknowledgement:
From 1992-96, when the poems in this book were written, the members of a poetry workshop in Halifax gave energetic help. I'm grateful to them, and to all other friends who questioned, suggested, praised, and prodded. Thanks also to the Banff Centre for the Arts, where the "tales" began to multiply in May 1993. Finally, for giving some of these poems—often in earlier drafts—their first airing in print, thanks to the editors of the following periodicals: *The Antigonish Review, Arc, Blue Moon Review, Event, Grain, The Malahat Review, The Officers' Quarterly, Poetry Canada Review, Pottersfield Portfolio, Quarry, The River Review, TickleAce.*

The translator of the quotation from Calvino is William Weaver; of those from Bachelard—Etienne Gilson and Daniel Russell.

Published in 1997 by:
Ekstasis Editions Canada Ltd.
Box 8474, Main Postal Outlet
Victoria, B.C. V8W 3S1

Ekstasis Editions
Box 571
Banff, Alberta T0L 0C0

Granite Erratics has been published with the assistance of a grant from the Canada Council and the Cultural Services Branch of British Columbia.

For Karen, the blessing I never expected

Contents

A voice means this: there is a living person, throat, chest, feelings, who sends into the air this voice, different from all other voices. A voice involves the throat, saliva, infancy, the patina of experienced life, the mind's intentions, the pleasure of giving personal form to sound waves.

Italo Calvino, *Under the Jaguar Sun*

I

THISTLE PLAIN AND TRYSTING TREE

Always

Somewhere a wolf spider dances on a white rock
shaking in fervent frenzy. Somewhere a crippled auk
tries to fly, kildeer mate in a soccer field, a shrike
shoves a warbler onto a thorn. At this moment
a woman watches a meteor, a child counts the seconds
between lightning and thunder, old men share
ale made from malt, hops, and Scottish water.
Always, during your day and during your night
black flies pierce human skin, rice-shoots
poke through earth, worms tunnel, a mother grazes
her infant's cheek with love for the first time.
Always, heat at the heart of a crematorium is reaching
its peak, and a queen bee drops dead into mud. Pick
any moment: a couple on a mountain breathe air too thin
for their lungs, but feel inexplicably at home,
while a couple in their garden breathe the smells
of two dozen species, and feel lifted
into an exotic place. Now, as I write, rain cascades
into a shrunken stream, foxes nip each other,
a rotting peach loses its last firmness. As you read,
a skirt falls to a bedroom floor, tires crush
a crawling animal, fingers press piano keys.
What happens during a pause in your talk
could keep you typing until your last breath. Always
a bullet leaves a gun, honey pours from a spoon.
Your brain is a mussel shell that will never hold the ocean.

Flying Crooked

for Don Domanski

A drummer flutters his hands in a microphone–
hummingbird jazz. All surfaces struck,
rubbed and brushed, he slaps his rib-cage.

I'd make such music, pull surprises
from surprises. In dense forest silence
a door creaked and I pictured a shack
battered by the seasons, caved in–
then nothing but a bare branch was nudged
by a rhythmical wind. A door swung,
no door swung. Or one cold morning
in a balcony-filled city of millions
someone all clownish floppy white
danced dervish-limbed across a flat roof
until I found it was only chimney smoke
wind-pushed, stretched and shredded.

To show you the fritillary's gilded design
      and its crooked-flying skills!
I won't take you to Sonny's Petting Zoo
where you can stroke a duck's head or tickle
a goat's chin, as if all were declawed,
stuffed with bloodless fabric.
                                    I'd shoot the rapids–
but not too fast. Who shoots them the *slowest,*
seeing whirls and thick salmon and flung froth
yet keeping aimed and afloat?

Would that my words bent down and drank water
from a moose track, made bookends
with iron from an abandoned bridge, crossed
the home of mudworms and mussels at low tide
to pick rusted lighters from the sea floor.
In a middle-of-nowhere town, from a bus window,
above a snowy creche, I saw angels of fish-nets.

To be fleshed with eloquence
like "poor as Job's turkey"
or the signs hammered onto roadside poles
BLIND CREST   OINK MART   NIGHT CRAWLERS
or the phrase on a faded map's pasture
thistle plain and trysting tree.

If I gave all that, you'd feel the freedom
that lightened my legs and shoulders years ago
when a welcome blizzard shut school down
and I walked home
with the gift of the whole afternoon in my arms,
struggling to breathe, wide-eyed to see,
whipped across my face by a whistling power.

Granite Erratics

To personify If—But—And....
Hawthorne, *The American Notebooks*

1

If has a tail shaped like a questionmark.
Among meshed branches, it floats from spruce
to pine, pine to birch. The playful one. Furred
with conditionals, its nimble feet twitch even in its sleep.
If *if* weren't with us, the brain might be little more
than a register, the sky a sealed roof.

What if anti-gravity always kept us an inch
off the earth–if frontier artists heard the truth:
    being painted shields you from bullets–
if the curator didn't exaggerate a bit
    when he said "A toad is in heaven
    with a breakfast of earwigs and June bugs"–
if the cat obsessed with reflections speeding up the wall
    leapt, and at its touch lightbeams became birds.

2

"But,"said the wounded, wide-eyed sailor, "but...,"
seeing the surgeon pull off his wig and lift out
his own glass eye.
        "But," said the seventy-year-old woman
buried under earthquake rubble, refusing to come out,
"but I'm naked, and people will say I have no shame."

But rears its scaly head, blinks. Now it hisses,
cantankerous; then, makes the mouth's subtlest upturn.
Alert, *But* finds the generous thief, the sour pit
in the sweetness, the lighthouse killing blinded birds.

Where would we be without *But*?–under a government
whose voices all huddle into one, on a planet
    without shadows or valleys.

3

And is *And* the greatest of these?
                                        Ever-migrant,
at home anywhere, winged and gilled. Eyes bloodshot
with smoke, sea-water, cave-air. Its diseases many, its beauties
contradictory. The generous one. Finds it hard to say *no.*

We have mosses that live 2000 years—a year
and a year and a year–and vireos that sing
20,000 times a day–a song and a song and a song.
We have birds' feet, forgotten by many,
poor cousins to feathers and wings–
                                                            a grackle's (black)
grasping a wire fence, and a guillemot's (red)
backpaddling, and a willet's (blue) crisscrossing sand.

And crosses the spaces between this and that.
On the coast, step among the granite erratics–
boulders pushed from other latitudes by glaciers
aeons ago, monoliths on wind-flattened plateaus–a rock
and a rock and a rock. Totem to nothing,
each claims a place
          among the haphazard many,
there, inescapable,
          immovable for the meantime.

Ask About Babel

for Mark and Sue

The librarian does not believe what she sees.
Her eyes are sad
and she walks with her hands in her dress.
Mark Strand, "Eating Poetry"

Here stands a building that houses enough books
    to fill a life multiplied many times,
to tell all strangers how little they know.
    SILENCE signs have been ripped from the walls.
Laugh from your belly until only a smudge
    is left of that picture in your head,
the grey-faced bun-haired bookworm
    who "begins to stamp her feet and weep,"
whose breath holds odourless dust.
    Be forewarned: librarians jest, curse, sweat, jostle.
Learn how lucky they are to work
    in the thick of it: planet as book, book as planet.
Ask about Babel, they'll turn the key
    of the elevator to its tower. Look harder:
they squeeze into bathyspheres and
    sink to regions sparsely charted, they climb
mountains with trails chaotic as botched
    bibliographies, the air bearable to few.
Whistle and cheer when they return
    with a flower that feeds on snow,
a diamond-shaped stingray whose tail jolts you awake.

A Sliver of Dawn

Did I open one eye, both, neither? When wind
knocked a blind's cord against the sill, a jay
shrieked in the garden. Ten seconds or so of waking,
then sleep pulled me back down by both legs.

I can't recover much except that voice
livid in the lilacs or the maple—a raw cheer,
                    an alarm, a life-blood call
that might've sprung from my own throat.
No chorus of ten species sends higher waves
through the air. I grasped sound like an outfielder leaping
to a high fly, then sank back into sleep
                    empty-handed.

The many-roomed cave of dreams gives us
other eyes and ears. Its fears end
with miraculous suddenness, unlike ones we live with
all day. But as far as I know, I've never heard a jay
in there. What I've not witnessed during sleep
could take years just to tell—mating blue-black
dragonflies rasping and vibrating
on a peony bush, an eclipse that made the air
a blurred furriness, an Indian rubber ball
streaking down a street with nobody in sight...

From sleep's deeper caves I might offer
a different report. But this morning
I felt closer to home when a jay
cracked the air, and for a few moments
—muddled, struggling—I wanted the day.

Poem on a Forgettable Morning

Who can count the ashes of past hours?
In the sprawling pit where the days go
most are extinct, extinguished, but stray winds
stir embers to the top, fan some mornings
into a glow:
                    say, a woman with animal-skins for clothes
and wool-thick hair to her waist
strikes two stones together—the first cooking-fire...
or a singer on a split porch
fooling around with a tune
scrapes a curve of broken glass along taut strings—
the dawning of bottleneck guitar.

What are ashes to some are embers to others.
I have my own burning ones, recoverable
or not: the day my ancestors loaded wagons high
with chairs, pots, silverware, bedding
and left Maine for New Brunswick, or the day
my father and mother first kissed full on the lips.

Memory—a hand too small to hold
more than a few charred sticks—lets most nights
fall from its grasp. Is there any buried ash
that doesn't deserve a word? Twenty-four hours
of pure forgettableness?
                Climbing down into the pit
where mornings and nights go, I would lift one
into the light, and say at least "It was."

The Stirring of Notes

for Karen

In my broken-spined jotting book
  old phrases slept for years, fragments
    curled into themselves:

délice de Bouddha/Buddha's delight from the menu
    in the window of a Japanese restaurant
    smiled emptily until tonight
    when it found a fullness in you

caterpillars everywhere—under a tree,
    thick in the air like thoughts of you
    falling onto my shirt sleeves

breathing suddenly heard on a string-quartet CD—
    how could I know that such release, breath
    exhaled when one violin paused
    before another resumed, would be you?

    even lone words—
hypaethral (roofless)––
    open into you, where other planets slip through
    like the night we sprawled on our backs
    in an island's grass and hawkweed

Tonight before you secretly made
your way to the desk, then lowered your arms
down onto my shoulders, your mouth
onto my ear,
            that notebook was a forest
where anything hibernating
awakened with your footfall.

In the Last Act

In a theatre with walls so thin they shook
when buses went by, the actors' lines were ropes
pulling us into the darkness under the lights.

Somewhere in the last act I heard a sound—
off-stage, outdoors, a barking like an echo
of something unsaid by the actors. What was it
they didn't know the dog outside knew?

A woman and a man tore from each other's embrace.
He waved a candlestick at her, she stumbled backward
and covered her eyes. Their story wouldn't let me go
yet that barking changed to wild godalmighty howling,
then back to barking, loud whimpers in between.

Her glove around a suitcase handle, her face
bordered with furs, she waited by train tracks
in falling snow. Her escape held me, but why
was I drawn even more to that creature, whatever
clawed at the sides of the evening, lost
in a dead-end alley or limping on three feet
or tied to a wire fence? In their scripted pauses
did those on stage want it muzzled or kicked?

Sweat broke out on their foreheads. Their voices
carried like struck brass, but soon
all I heard was stillness outside.

Postcard from an Unknown Friend

President of the Fifth Court of Hades
Lead-glazed stoneware sculpture
Ming dynasty, 16th century

An image to give you an appetite
for more than soft bread and porridge, to tack

beside your mirror and study your character by
after sleep: his stiffened hair pulled upright,

skin green-streaked mustard, ear-lobe bolt-holes
empty. If some benefactor mailed this face

to every address in the country, sluggish mornings would grow
lighter-footed, nights shed their numbness. Fiercely,

fully itself, it can be ignored like a rock on your plate,
a thorn in your eyelid, an octopus twisting

under your sheets. Locked to those eyes—
black-yolked eggs, warped planets—feel your fears

gather at the back of your brain like moths
lightly brushing a window, unnerving *because* they brush

so lightly. Keep this grimacing ancient
brother of us all. He can help you step up

from one stair of waking to another.

Massacre Jigsaw Puzzle

1

On top of nutcrackers and pine boughs
    printed into a festive tablecloth
long-helmeted soldiers on horses
    aim spears taller than themselves
skyward. Iron armour covers their mouths.
    In a season named for a child's birth
we piece together the deaths of many.

Upstairs the baby sleeps—nephew,
    grandson, son. A monitor microphone
links his crib to us—a cord
    finer than the eye can see
strung from some star to the earth.

2

Prayerful hands tight together,
    a townsman on his knees pleads
with a stranger in royal red, the horse's tail
    braided in prim circles. A soldier
tugging a boy by the hand
    half faces the startled parents
as if to say *Don't worry, don't fret*—
    Uncle Pieter taking his nephew skating.

On the monitor, a sound between a sigh
    and a groan, a kicked slat.
Blood must be on the ground, splashing
    horses' fetlocks and soldiers' boots,
sprinkling mothers' faces and hair.
    Who gave us this present?
What happened to framed embroidery

Journeys of the Magi, paint-by-number
Adorations of the Shepherds?
We could heave this, box and all,
into the fireplace, but we're driven
to find a mother or father for each
so no one is stranded on a fragment
when the whistling sword finds its mark.
Also—admit it—curious as tabloid
addicts, we want the panorama complete,
every hoof print a bruise in snow
until the room whirls with screams and dust.

From that star overhead, a whimpering....

3

Some whitebeard—judge, rabbi?--
snaps into place...leading the soldiers,
no Solomon in the end. If we could see
all the tentacles of Herod's power
curled around necks, they'd make a thicket
no lamb could enter.
A collapsed mother's naked infant lies
dead in her lap like a figure of snow
half melted, then frozen again.

Through our windows night falls
into the jaws of a toothless cloth dog,
the jug of punch, its floating bits of fruit.
The whimper becomes a cry, the invisible
cord shakes, the puzzle locks us in.
When will one of us rush to the stairs?

Skylight

The day my cousin shot himself in the head
I was cleaning the skylight in the hall.

The landlady's loose-nailed stepladder
shook with each step I climbed down.
Over the phone my father's voice, 500 miles
away: after a furious standstill
with his wife, my cousin had carried
his cherished inheritance—our grandfather's rifle—
past his sleeping daughter into the back yard.

At the top of the ladder again, I lifted out
the glass. Dead flies, grit, rain smears.
His mother was saying *never talked...*
like a shut book.
In our grandfather's hands the rifle
had provided deer meat for his father.

I took the glass to the bathtub for a wash,
jets of Windex, a speedy polishing.
Rare games of crokinole and baseball
had built a flimsy bridge between us.
After its collapse, the time he tried college
I'd glimpse him in the Student Union
shooting pool, game after game. One night
a decade before the end, he was a waiter
in a bar, serving me a beer, silently.

I climbed the ladder a last time to lift
the glass back into place. The hallway
was like a new world. For the first time
I could see sky through the ceiling.
How do we deserve such an inrush,
such clear watery light?

The Uprooted Sleuth

A cigar tossed from a train window hits a cyclist in the eye.
A plot might start there. The sleuth picks up and sniffs
the possible clue. Sounds of his brain at work
drown out the trilling of the toads in the ditch.

He misses the city's chimneys breathing out secrets,
its doorways scarred with crimes. He is ten disguises,
                    a genius of stains, a blade
honed to one purpose. A rook means nothing
deepening its wing-beats above a steeple
unless a bullet in its breast leads to a man
circled in guilt. A spider web crossing a shut gate
must prove something: nobody went through.
Moors must be a stage for tragedies,
                    a site for burials.
Why a woman refused to cut her fingernails
intrigues him, but a grebe's fleet mating dance
just beyond the boat in which he scowls
skitters past as far from his mind
as a waterwheel turning in northern China.

When he finds no motives to untangle
like the heap of oily rope curled at his feet,
boredom creeps into his cells.
                    Lonely as a lost moorhen
whose head cannot stop bobbing, he holds a pocket glass
over the cracks and grit speckling his boots. How hard
he hunts for something to make his heart speed up.
Around him trees sink into their own reflections,
chaffinches sing from the hedges of another world.

Needy Space

...all really inhabited space bears
the essence of the notion of home.
Bachelard, *The Poetics of Space*

He wants to teach classes under the sky.
Hot as a hospital ward, L-201
has all the character of a stripped soup can,
cold tin. Why are the walls blank but for a clock?
In elementary days, paintings made the walls
move—a train with many-faced windows
curving through mountains, a sunflower showering
gold into his eyes. Is a windowless room
any place for learning? Give him a forest clearing,
tree stumps for chairs; a wave-chiselled cove,
air streaming over water and everyone's face.
Does the fool think he's some dryad's son
torn from his native place? *Okay, bow*
to monkish ways, face the faces, voices
and minds that make a place what it is--
Joe shaping answers in the back row,
Jenny in the front. This class has more kick
than any he's coached before—but the room
shaves the five senses, and he secretly
grieves for it. And nothing
he does there is more than a potted
cactus balanced on a podium.

The Woods On The Way To School

for Mark Abley

When well sheltered, memories are reborn
as rays of being rather than as frozen shapes.
Bachelard, *The Poetics of Reverie*

A globe in the classroom window marked the end,
  the turning indoors for the morning,
    the afternoon—blue bulb with pale continents

fixed, spied through glass as he rushed
  to the doors, panting, late, his delicious dawdling
    finished. While he leapt over logs and followed

the tricky shadowed path hidden from all houses
  he'd hear the bell clearing the jungle gym,
    pulling him away from the spongy-mossed

ground, the catbird expanding its mimicry,
  the toadstools too flesh-like to touch. Late,
    late. Tardiness x's like birdtracks

crossed the boxes of his attendance record,
  badges of virtue in *his* eyes.
    Through the four seasons of grade four

each morning he walked half a mile downhill
  past shrubs and chimneys and fences,
    then let a thin ragged strip of woods

—a short-cut that slowed him down—swallow him up.
  Full of perplexing turns, the path was followed
    by strangers he never saw. Thirty years later

I find nothing but a ghost of those woods, new streets
in its lap, yet an impossibly familiar tree
still stands, and from it I try to build

the old space—like staring at a time-bleached
circus poster on a half-collapsed barn
until the field flickers with lions and clowns.

How narrow and bounded it all
must've been, how slight a break
from marigolds and carports, but for a boy of ten

those trees opened into a wilderness
rampant with secrets—a dog's skull driftwood clean,
a clutch of red-necked-snake babies writhing.

When I read *sylva sacrosancta*
I think, yes, for one year that was my holy wood.
Swaying between two times, I sit high in a tree

climbed by nobody else, watching the thin boy
I once was. It's no fairy tale, but a story
imprinted in my brain cells deeper than I can go,

in the lines of my hand held up to my face.
Many days the boy took the trail four times—
after breakfast, back home at lunch, schoolward

again, homeward at 3 o'clock—so he picked
his way in all sorts of light, milky, lemony,
purple. For a month he didn't know

all he was missing, then his first pair of glasses
brought distant dimnesses closer—birchbark
with staring slit eyes, a raven's throat

wind-ruffled, snow turning skinny saplings
into immaculate stalagmites. At the dinner table
his father said, "If we saw like hawks

we could read headlines a quarter-mile away."
For days he dreamed of having a hawk's eyes
transplanted into his face, throwing

his glasses away to become some Super Hero
gifted with the strongest vision of any boy on earth.
Years would grow out of years

before he knew: poisonous mists and liquids
blurred warblers' voices and loosened
their feathers' roots, sea turtles mistook

collapsed balloons for jellyfish, swallowed them
and suffocated. He was cocooned by ignorance,
an envelope lined with ancient moss.

One morning a ray of light like a Maliseet arrow
hit the bull's eye on a flat stump—
a huge beetle's back. He'd never seen

such luminous blue, like a drop of congealed oil,
a bit of a stained-glass Jesus robe.
A fire in his throat woke him that night,

by Sunday he was in the hospital surfacing
from ether oblivion, his tonsils
a frog with a belly full of live coals

flushed down some chute. Bored by the blank
walls, sore-throated, thirsty, he burrowed
into the get-well present from his mother,

a shiny-spined book on beetles. What chords
it struck, what music in the names
among the quarter of a million kinds—

kangaroo beetle   cucumber beetle
Hercules beetle   harlequin beetle
ambrosia beetle   striped blister beetle

rose snout beetle   buffalo-carpet beetle
  Mexican bean beetle   pleasing fungus beetle
    confused flour beetle   twice-stabbed lady beetle

tumbling flower beetle   six-spotted green tiger beetle
  whirligig beetle   devil's coach horses
    —but he's held on to no beautiful facts

except ones touching on sexton beetles
  who could've spurred him into praise
    translated: *They do what worms cannot do*

and the beaks of scavengers decline. They work
  for free, send out no catalogues featuring
    mahogany gold-handled coffins. En masse

they can bury a mouse in two or three hours,
  a rat in five or six. Their own limbs and shells
    serving as tools, they plant their eggs

in the flesh of the dead.
  In the hospital room hot as Brazil
    but bare as an Arctic waste, he didn't ask

what I ask, how much the names—*the names*—
  magnetized him. The other day I savoured
    reading *peppered moon lichen, lettuce lung*

lichen, dimpled specklebelly, cryptic paw lichen...
  I remember now, but even then memory
    was a garden or a briar-patch forever crowding

side vision. I figure memory must've started
  right after the cradle, even if later chapters
    cancelled the first ones, like weeds

pushing out weeds. At ten, the boy looked back
  at palm trees he'd made rolling up newspapers and
    slitting them from the top, then spraypainting them

silver or green; he uprooted hours when leg aches
  woke him into the middle-of-the-night misery
    of growing too fast. So I don't laugh too long

when the radio interviews ten-year-olds
  opening their own museum. I see their footprints
    reverting down a road, the sizes shrinking.

One girl got nose-bleeds every second week or so,
  red soaking the teacher's tissues. The boy
    kept watching on the sly—unsure why—

the blood astonishing among desks and scribblers
  like his own blood the afternoon he ran
    into a rocky-bottomed lake and hopped ashore

with a red stream gushing down his white foot.
  In a country hospital while a doctor stitched
    the wounded flesh, the boy couldn't forget

the sweet sensation unwinding within his pain,
  the woozy satisfaction knowing his body stored
    such thick brightness. Sex was still an unknown

aurora, a river bed waiting for water
  to rush in from its brooks, while shapes and sounds
    seduced him everywhere—salamander eggs

like jellied packets of eyes, apple blossoms
  fallen on black mud, a wren's frantic flurried
    meandering higher and higher

up the scale. His body a register,
  a taken-for-granted set of clothes. I wonder
    if he ever looked in the mirror then

a year before a girl touched the curls on his forehead
  leaving a sensation subtle as a light beam
    but strong as if a doe had struck him

with her hoof. Even the time a hard snowball
  shattered between his legs and pain
    rose to his ears and lingered all the way home,

he never thought to look for a bruise, his body
  retreating like a shadow when clouds
    slide over the sun. It was a year when humans

shuffled in the background, signifying less
  than things—as if on a stage
    the painted trees and bushes walked forward

to wave their branches and talk, and the actors
  withdrew into the wings. Yet I know
    differently, I take a spade and unearth

a few faces from the murk...
  Everett from down the street disappeared
    one spring night, but by dinnertime the next day

he was back, and neighbours knew he'd hidden
  in the woods, shamed by his report card,
    afraid of his father's wrath. Why did the boy

envy Everett? He was tempted to ask
  how those hours had unfolded—spots of green
    phosphorescence, the very breath of the woods

like a brother—but when thunderclouds soon threw
  down darkness large as a plague from Exodus
    he felt cast into that night: sleepless

wrestling, clammy cold seeping through his shirt,
  the guilty document like a bloody knife
    in his bookbag, the woods the farthest thing

from home he'd ever known. Then, the day he flung
  a clump of dried mud at a Stop sign
    to test his arm, or for no more reason

than the pure pleasure of doing it, and a honk
  behind his shoulder announced Sergeant Lawson leaning
    out a paddy-wagon's window to scold and warn,

the lungs of authority hoping to make the small
  quake. Such a thin moment, such a speck of dust,
    yet leaving a fine clear scar:

in a bar a thousand miles from that street corner
  I berate myself for saying *Sorry*
    as if all my cowardly acts since have denied

the delight of watching the mud explode
  like a meteor striking a planet. And I still see—
    faint as smoke just before it disappears—

the face a girl named Clare brought back to class
  days after her father had jumped off the city bridge
    carrying his debts and his unreachable secrets,

the grief in her eyes like nothing in the woods....
  With a pencil the boy drew his brain as a factory
    spewing out paper, an insect-eyed worker stabbing

two dials at once, cogs and wheels turning
  everywhere. Why didn't he sketch
    his brain as a forest? Was machinery

closer to the truth?—his thoughts inside looking out
  through splotched windows, not outside
    swooping and kicking with the birds and squirrels.

Fingerpainting gripped him
  like nothing else in class, he flung himself
    into trays of many colours, all that

unpremediated play shaping lines and ovals
  like branches, roots, and leaves changed
    beyond recognition by the twisting

winds of his feelings. At supper he'd see paint flecks
under his fingernails like earth clawed
from the tangled floor of the woods.

A man walking miles from any house or road
picks up an oak leaf fallen from its twig
and sees in its veins and clinging wet

a long pocket of trees from thirty years ago,
the pocket with a hole that emptied him
into a schoolyard, above a hard-to-believe

globe in a window. Through every forest he finds
the trail that followed him from his tenth year
into his eleventh. In a foot-sized pool

rimmed with cones and pods, in the depths
of a wasp nest like an abandoned clay headdress,
through a bullet hole in a *No Trespassing* sign

he sees the year he first knew he was
spellbound, when he detoured from the street
into a land of blue beetles and moss.

II

A HANDFUL OF TALES

He looked into the water and saw that it was made up of a thousand thousand thousand and one different currents, each one a different colour, weaving in and out of one another like a liquid tapestry of breathtaking complexity; and Iff explained that these were the Streams of Story, that each coloured strand represented and contained a single tale.

Salman Rushdie, *Haroun and the Sea of Stories*

A Basement Tale

for Andrew

1

Twin brothers slept on thin cots near a furnace.
One heard a gruff lullaby, a fiery-hearted father;
the other, a smoky-fingered devil sharpening a blade.

With a click and a shudder, the furnace shut off.
For one boy, silence was a shoreless pool rippled
by a single minnow; for the other, a spider,
the tips of eight legs pressed against his face.

Rain slapped the basement windows and seeped in
through cracks—the sky at work feeding rhubarb
in the back yard; the slow growth of mould
over split stones in a Loyalist burial-ground.

When the bathroom fan overhead whirred, one brother
heard a whale, fathoms deep, serenading its mate;
the other, an open-mouthed ghost trapped in an iceberg.

2

A click, a shudder—the droning began again.
The twins crawled from bed and slept in the furnace.

One walked into a tropical forest hung heavy with fruit,
shot through with birds spanning the spectrum;
one fell into an inferno that burned up his clothes
and glasses, his eyelashes and lips.

Each brother wrote a book—one with berry juice
and feathers, one with his finger stone-slashed.

3

Upstairs the next day they read each other's words,
baffled. Baffled, curled in back-to-back chairs,
they knit their brows into mazes without threads.

That midnight, back on their hands and knees
they crawled down a trail toward each other's dream
from the crossroads inside the blazing mouth.

An Imp Tale

1

Twin sisters were told never to laugh in church.
Phlegm-throated pigeons on the roof
burbled some service of their own, a young deacon
smelled like he swam in pools of after-shave,
an old woman slurped her gold tooth. But the twins
locked up the imps of laughter—their father's words
like the vaccination scars on their freckled arms.

They fiddled with the hat strings at their throats.

2

That Sunday an out-of-town choir filled the loft,
robed in orange. One sister rolled her eyes, whispered
pumpkins, and her twin coughed. Hunting for high C
a soloist's voice wobbled like that of a lady
slung over a villain's shoulder, shrieking, kicking.

When the silver collection plate slipped, a quarter
raucously rolled across the floor. Their father
grabbed the plate's edge, but not before imps
jumped from their throats. They slapped their hands
to their mouths, to keep the others inside
while the soloist dove into alto.

Who'd tipped the plate? Each was sure she had.
But one felt free; the other, baptized in guilt.

3

During the sermon one sister watched her urchin scurry
like a tiny monkey down the aisle, up the pulpit,
over the gold cross; her memory blurred the fleeing coin
into a rollerskating track down which she sped,
monkeys clinging to her ankles and neck.

Her twin ran over the backs of pews sweeping a net—
the rolling coin a road down which she lurched
dragging a wagon of shame behind her in the dust.

Before the morning of the pumpkin-robed choir
the sisters' straws had shared the same glass.
On the way out, joining hands, they saw their imps
claw and spit at each other in mid-air.

A Vacation Tale

1

In a small house where farmhands had slept
long ago, by a small sink in a small kitchen
a crying woman washed breakfast dishes.

Her children were three radios, station dials
spinning all at once: the girl jumped
into and out of the wood-box, the boy banged
a muddy shovel against the floor, the baby
tore the bars from the crib.
She raised her dripping hands, lowered them.

From a broken pump-organ, a mouse-gnawed Bible
toppled open to the plagues of Egypt.
When she stitched herself a tunic out of rags
nobody noticed it hanging from her shoulders,
soaking up the sweat down her back.

2

Where was the husband?—in the big-gabled house
up the hill, talking the days away with his father.
In deep leather chairs parked by footstools
the two men chewed over taxes and the past,
a jug of lemonade resting on a tray
decorated with Venetian gondolas.

Napping, the husband became the child who tied
his many brothers' suspenders into a lasso,
then spun the house around and settled it
back down on a nest laced with ancestors' hair.

3

Each year before the month ended, she climbed
the hill up into the grand verandah:
in a gondola the two men smoked cigars, a sunset
smeared over water, lute music falling from balconies.

She stood on the bank, called and called.
From a basket at her side, she offered them gifts:
two logs, two dirty diapers, two broken-backed mice.

4

Heat drives the children into sleep, the husband's yawn
surrounds the moon. In the rear-view mirror
all his brothers and sisters chase him barefoot
down the road, and his dead mother holds high
a loaf of fresh bread, a jug of molasses.

Home is several slow hours away, but the wife
gazes ahead as if she sees it—a star, a planet.
With one hand she kneads the tunic of rags
balled up in her purse. If she throws it out
the window, will it flap away, a fleabitten bird
swallowed up in the wind?

A Bear-Hunt Tale

1

A shivering father and a shivering son climbed high
into a pine tree and waited–bows and arrows across
their knees, a rust-freckled barrel of bait on the ground.

Days after drinking his thirteenth-birthday rum, the son
echoed the father's curses, nursed the father's tobacco
in his cheek. As dawn dwindled, he propped his head
against the trunk, as if the bark were feathers.

2

Carried off to sleep by a mother crow
he found himself inside that barrel, battering
his way out. He swung along the beef scent–a jungle boy
along a vine–to a brook where a she-bear soaked
her flanks, the scent threading through her nostrils
like irresistible musk from the male's gland.

Choking on tobacco, grunting awake,
he spat a hot stream. He wanted it to start a fire
beautiful as his uncle's red bowling shirt.
Why didn't he breathe a word about such flames
to his father? Like big tap and little tap
they pissed together over the edge;
the father was mighty water pounding down a canyon.

3

To keep the jaws of boredom wide, he figured how big
the barrel was: big enough to trap and nail his sister in...
to hold all the rum for the wildest barn-dance
heard by horses miles away...to build a raft
for flying through rapids, over drowned faces...

As the list grew, he kept it to himself
like a fish thrashing under his shirt.

4

The son wasn't listening when the father whispered.
A high-pitched sigh ripped air, an arrow sped
through flesh, a rasping bear clawed earth.

On the way home, humming a tune, the father was sure
the son smiled because such a beast, all rage
and teeth, stiffened in the back of the truck—
but the son smiled because the barrel could make
armour for a bear to wear in battle.

A Drum Tale

1

One winter night a boy became a babysitter
in a bungalow seized and shaken by wind.
Who were they, those neighbours so little like
his parents? In the man's fur-trimmed slippers
he flung his Physics text aside and roamed:

nudes on shower curtains, lascivious-finned
tropical fish, *Provocative Percussion*—
on the record jacket, a great-bellied woman.

Shaking, he slipped the music from its sleeve
by a clay dwarf's lidless eyes and shovel-like penis.

2

In the chimney turbulent snow moaned
You've never seen the sleeper awake...awake!
If the baby kicked the wall and wailed
would none of his faltering words chase away
its nightmare? What if it looked *too* asleep,
still as one of the smothered princes
in his father's brittle-paged *Books of Knowledge,*
in the engraving all cold dark blue?

Don't wake...don't wake...
The elegant curved stereo-arm zigzagged down.

3

He raised his hands, loosened his hips, leapt
between the record-jacket woman's legs
past waterfalls and cliffs to a clearing

where girls lightly skirted in straw danced
his way, closer, closer, imprinting their bare feet
with blossoms and seeds. He muttered *can't stop*
moving can't stop moving—mating time,
his mating dance could win him a wife.

Sweat falling from him like smashed jewelry,
he touched a smooth honeysuckle breast.

4

From the bedroom with the closed door
a voice rose, plaintive, breaking. In the hallway
he still felt an unreined rhythm charging
through him, drums in the background,

but over the blizzard pulling at the eaves
he heard a baby—his own—crying out.
The father in him, fine-veined and afloat, turned.

A Cave Tale

Venus Meets Vulcan

Sigismund Goetze, 1909

1

The gallery was so quiet the paintings murmured,
an old commissionaire relaxed bone by bone.
Across wax-brightened tiles a girl's sandals
squeaked, goose bumps over her bare legs.

Gentlemen with guns, dogs and dead pheasants,
pale-cheeked children buttoned to the neck
and gowned ladies against swirled skies
gazed out with eyes indifferent as plates.

When the girl stepped up to another world—
vague thick-muscled men in a cave, a woman
light-stroked at the stone mouth—
                                    she gasped
so loud the slumped commissionaire stirred.

2

Eating walls and distances, the dark
had always frightened her—a closet her brother
locked her into, laughing; a parched well
stealing her voice; streets after school dances,
all the boys twisty eels or furtive moles.

Three-quarters of the canvas was layered tar,
swallower of colours, a night inside the day.
In the upper right, the rock gave way to seaside
silver and green, a luminous-lemon tree.
Dizzy, she saw man and woman turning
toward each other,
          the lifted hand of Venus
like no greeting she'd seen—love no longer

a parasite on sunlight, but a brave self
drawn to a forge burning orange, one foot

on the first step leading down.

3

The ladies' and gentlemen's plate eyes
cracked, uncovering cliffs, flames, hammers.
Venus and Vulcan burning into her back, the girl
fled the air-conditioned chill and settled
into a shaded park bench.
                    Unmoving, she climbed
down an opening carved in rough mineral.

She wasn't light to the darkness.
The blacksmith in there was no boy she knew
but herself facing a lit figure coming closer.

A Best-Man Tale

1

A best-man forgot the gold studs for his collar.
Fallen to his knees at his brother's feet
he begged forgiveness, then gazed up:
two tiny nails driven into his brother's neck.

While the ushers barked foul names in his ear
his father, scowling, choked him in a headlock
and twisted gold foil through the collar slots.

2

To the best-man all ceremony was death.
The box holding the bride's ring catapulted
from his bony fingers. Later when his cuff
bumped a brimming glass of wine, a red tide rushed
over the shoes of the three-hundred guests.

The emcee blared, the heat mounted, video cameras
swivelled—extraterrestrial eyes. His neck swollen
from feasting, the best-man obeyed the order
"waltz with the maid-of-honour."
Who manacled anchors to his feet?
In his arms, the bride's sister shrank with shame.

3

His Liz or Lou was lost in the crowd.
In their hotel room the night before
they'd rerun old skirmishes, slept without touching.
In the dance-floor swarm of elbows and legs
he found her and smiled. She slapped his face,
the sound a skin drum's stray beat.

Weeping, wandering blocks through the humid night
he knotted his burgundy tie around a lamp post,
waded waist-deep into a fountain pool,
threw the gold studs toward a stone nymph

who took on flesh and opened
her lily-scented arms. The best-man answered
yes, but only if the celebrations
stayed small—no fanfare
but nighthawk cries falling from high above
and a few hidden insects rubbing legs.

A Puffball Tale

1

Every game but one vanished from the Common.
Caught in a wide mesh of fog, a batter
wavered in and out of sight like a sailboat
nudged to and fro, nine-tenths invisible.

In left field a young widower kneaded a glove
with his fist, tweaked the peak of his cap—
a bird-crested cap he'd worn in bed with his love
one night when laughter shook them to the root.
After her drowning, his hair had turned white—

memory, a tree he hoped to chop down and burn
to its end.

2

Like an axe thunked into wood, the well-scarred bat
sounded. The ball, climbing, became one with the fog.
While the widower's gaze swung back and forth
the fog rolled on like a cloud of spores.

Puffballs by a forest trail—he'd picked her one,
pressed it between fingers, the spores spreading
a tiny volcanic surprise. He and she sank
together under a mountain ash, its fruit
clumped red past her kiss-bruised shoulder.

She was there, above, wide-armed in the swirl,
then his sneakers slid two ways and as he hit
the slick grass he tasted his loss,
a sickness no drink could wash away.

3

For months the sun was no friend to him.
Whenever fences and docked ships disappeared
he roamed the streets through his truest element,
that smoke-thick air, awaiting the moment
the wetness would open into her face again.

Under any merciless blue sky, grief
turned the light into a gallery of knives—
those rays carving clear shapes everywhere.

A Swimmer Tale

for Liane Heller

1

Each morning one winter a woman swam in the sea.
Strangers at their windows wondered,
stories whirled thick and devious as a blizzard:

she was spotted in a supermarket buying squid,
at bingo wearing starfish ear-rings,
a dolphin brooch. Did she only dogpaddle
near shore, or ride some sea creature's back
out to a rookery island? She did it
to build her muscles, or make the city whisper
about her, or wash herself clean of deadlines.

When a boy who called his dog Astronomy
swore she was hairier than a man, her father
a harbour seal, his mother broke his telescope.

2

From her head to her feet, blood throbbed
like lava. Pain turned into the fullest waking
she'd ever known. A ship's whistle was her voice
springing up to the swollen sky.

At the office she survived long slow days
unblinking at a computer, where what saved her
from boredom's bottomless mud
surfaced between the letters on the screen:
cresting waves, skimming eiders, cavorting seals.

3

Drawn to the hypnotic window facing the shore,
the boy begged his mother to look closer.
She gazed and gazed until she changed, then
lifted her brush to start a watercolour:
a woman's aquatic ballet with a tortoise.

Crystal teeth raking the swimmer's back,
the boy joined her in the cold. When she cried
"Help, help me," he turned and quickly thrashed ashore—
he'd wanted her ecstatic and brave, beyond need.

That morning her dwindling strokes took her
past the rookery island until the sea
slipped rings around all her fingers.

A Landlord Tale

1

A landlord thought himself a spiritual man.
Ten houses on ten streets—five dirty white,
four dirty blue, one dirty green and pink—
begged for fresh paint and countless repairs.
But the landlord thought himself a spiritual man.

In a house no more splendid than the others
he lived alone with a half-dead poplar bent over
on his lawn. When he poured whole mornings
into his violin, the world shrank
to his soul making no music but his own.

Badgered by tenants, he tracked down handymen
willing to work for the paltriest pay—
a thrice-convicted thief, a half-blind
jack-of-all-trades, a one-armed plumber.

2

A spectacular ten-pointed fork of lightning
set the roofs of his inheritance afire.
For a week the landlord wore black bands,
five on each arm. He ate nothing
but beef broth, wept for the deaths of so many
strangers. The sickly tree leaned farther.

Debts everywhere, he caught a ferry to another
country, the highway south a singular tunnel
through which he sped, oblivious
to motels and billboards crying for attention.

When his car collapsed, he bequeathed it
to the ditch, an abandoned suit of armour.

3

Near a village he rented a hut by a beach,
the bed and table legless. He was improvising
on his violin when waves yielded strangers
dressed in smoke stains. "We paid our rent,"
said a woman heading the half-familiar parade,
"we walked out of the fire and hid in the ocean."

Why for the first time was the landlord ready
for head-spinning punch, balloon-bursting, shouts
like fireworks? Why did he wish they'd all
join hands in a great circle turning on the sand?

A boy pulled the violin out of his grasp
and threw it to the mouths of bearded waves.
The landlord nursed his wrenched hands and watched
the naked tenants walk away over the dunes.

A Skater Tale

1

A blind man in despair walked with his dog.
Not even sherry like a swallowed flame,
a many-rivered sonata, or his braille
Bullfinch's Mythology cheered him any more.

From a snowy footbridge a whispery rhythm
slowly grew closer, sharper, a skater circling
a small canal. On the island in the oval
the blind man felt no red-cheeked Mercury darting down
his body's frozen streams,

its winged feet snarls of bone and dust
under the raised script's hard snows.

2

What wild whim did he give in to, following
his dog up the steps of Apollo Sports
where a girl with a peach-soft voice
searched for skates his size? At his back
he sensed skis awaiting fields and woods,
toboggans that had never touched a hill.

Over thick socks, the good fit of new skates
startled him.
                    Flight soared to his mind,
promising paths that start at the feet and stretch away,
no horizon waiting to knock him flat, endless space ahead.

3

At mid-day the mid-city rink lay silent.
He halted, he wavered, he saw himself
a buttoned-eyed scarecrow blown off its perch—
but soon his shoulders and legs recalled it all,
his humming muscles retrieved their buried speeds.
Flying   free   hawk-sure   exultant—

When a fissure tripped him, he staggered
backwards, keeping his head high. Fallen,
he was a spinning X, his bruised elbow a small price.

His dog licked his face with its hot tongue.
It was the sun grown fleshly, telling him
the cradle of old myths swung in his gasping breath.
In his ears, other children of Mercury gathered.

III

BRIMMING

Brimming

From nine letters and landscapes

London

Are you sleeping across my most treasured hollow,
my side of our bed, under the burgundy duvet?
May Day in mayhem city, where I keep subtracting
four hours: in this B & B breakfast smells
climb the stairs, but in your time night's still
got a long way to go.

Into an envelope I slip
the burnished black feather of a Tower Raven.
When I picked it off the royal grass
no Yeoman Ravenmaster bellowed and charged, swinging
his halbert. Believe me, this feather's worth
one of the 28,000 jewels in the Imperial State Crown!
Drop it in that hollow I long for and belong to.

Prince Hal's alive: on the Underground late last night
in a circle of laughter-haloed friends, he crowed
"We're all individuals!"
(sounded like *We're owl indy-vee-jewels!*)—
a woozy bedraggled corkscrew-haired Falstaff's lad
knighting a day overflowing with wit-spawning talk.

But along streets beyond the Tower walls
among strangers slumped against brick—
blistered shoes, blankets over clenched shoulders,
heads bowed in sleeplessness or shame—
a girl burned me with her glance:
what scabbed hands, what grieving-cave eyes.

*

Whitby (North Yorkshire)

Along the North Sea they call clinging fog
sea-fret. When I'm home let's say "sea-fret"
while Cape Split's pinnacle withdraws, disappears
and reappears in the thickening, thinning air.

A sail-ripper wind batters the bench I shiver on.
Near tar-black tombstones stained by sea air
a sandstone cross praises oxherd Caedmon.
Around a table laden with tankards and trays
whenever the harp came his way, he'd carry a lamp
to the stables. "It is because I cannot sing,"
he explained in a dream, "that I left the feast,"
so an angel's voice touched his ear-drum:
"Sing about the creation of all things."

No harpist or singer, I'd raven-croak for you
about today's hike, six miles to Robin Hood's Bay
and six back: sea-floor rock bared at low tide
like deeply groved tabletops for crabs—
fenceless cliffs—wide fields of rapeseed, buttery
explosions—bluebells like pigment injected
high over a curving untrodden shore.

*

Rievaulx Abbey, near Helmsley (North Yorkshire)

Imagine living where the ruins of abbeys
                                        are second nature,
so we wouldn't give a second glance, as if they were—
in our world—a dead skunk or porcupine on the road.
Good to be a greenhorn, so your eyes feel newborn
                                        and things rush in
unfamiliar, freshly minted, however ancient—
even if, they say, by 1630
                    monument-lovers flocked
so thickly satire turned them into targets.

No ghosts of twelfth-century Cistercian monks
pray on their knees or mend fraying robes.
Dandelions lift their heads throughout the shrine,
red flowers curl from infirmary walls,
green shoots vie with rock in the sacristy—
as rivetting as shrubs and grasses feeding
off deadly basalt cliffs by the sea back home.

Though Latin and Greek once ruled the abbey
I'm besotted by *English*   by its stew   sponginess
ambidexterity   leapfrog   unkingly hoopla
and hijinx   its gut-rubbing and sky-tickling.

At lunch under flying buttresses
      I pulled a book from my knapsack and read
love must be moistened and bedewed with pleasures
(not St. Ailred, but Keats to Fanny Brawne)—
words that wheeled out into the sunlight
like lapwings, sharp wings dodging crumbled stone.

Missing you more than a monk misses his prayers,
more than ruins miss voices that once
                                  made them a home...

*

York (North Yorkshire)

All pubs are closed, thanks to a soccer match—
last year ale-soaked top-heavy fans
          broke windows, arms, pates.

With swords and cannonballs labelled and tamed,
today I tired of history shrunken
          to a glass case of knicknacks
and wanted fountains of blood to burst from under the floor
          spraying us all in the face.
Saxons, Normans, Danes, Norse—spurred cocks
lashing each other, the watchers at the pit
ear-splitting, spittle-lipped. A boot in the jaw
would've told me as much about origins
as a pub's chalked menu with its open-armed words
SIT AND EAT WHEREVER YOU WANT
(but reading that, I confess, cheered me to the core).

Meanwhile, around this bench by the River Foss
the air's charged, ripe. I'm close to brimming.
A preposterous-feathered peacock shrieks, a dog's voice
chases another dog's voice, a long slim boat slides
past with four rowers slipping oars in and out,
its creaks and splashes meshing with a tree shuddering
and an insect cruising around my ear. I'm brimming.

So far, no sign the four rowers will warp their oars
          into bludgeons and climb ashore.
What if you and I had been born by this river
to the tune of Yorkminster bells in our mothers' ears?
Another life for us appears like one of those boats
arrowing past, the two of us rowing, oblivious
          to the riverbank, out of ear-shot, gone.

*

Buxton (Derbyshire)

From "England's highest market town,"
Spa of Blue Waters,
Source of Blue Cheeses,
Aqua Arnementiae:

In Poole's Cavern the river—elsewhere a metre deep—
seeps over stone, a mere trickle.
Who would've called it *river*
if the guide hadn't said it was so?
From thermal springs, pale blue waters pulse up
through limestone. Gargoylish beauty
everywhere: stalagmites and stalactites
travellers' eyes have turned
into eggs and bacon, lions and apes.
Like the moors, they made me feel
I'd suddenly found a quirk in a parent
I thought I knew so well, a hidden
side of character revolved outward now.
To watch an ongoing drip-drop
build a cave's shapes—that's a mission
for a meditator, rapt, cured
of all restlessness, minerals mounting
a grain of a grain at a time.

We answer the call to travel but swear
we could shut it out, pulled
by no need to enter cities and cross rivers
and stoop through cave mouths
for the first time. Much of you's still secret
to me, but how much longer
before you're my happiest marriage of home
and away, hearth-lover and peregrine?
—and I that for you, your garden and wilderness,
your mole's burrow and roistering river?

*

Buxton

Stone walls, stone walls everywhere, stone walls marking
ownership and time. Crossing stiles, when I lifted
a latch, shut a latch, the unspoken trust warmed
chilled fingers, civility compact in the act.
But I blundered into an impasse:

a tussocky field closed in by stone walls
          topped with barbed-wire—
sharp little twists like withered hearts
intent on keeping others out or in, each point
like the knife-tip of a hermit barely sane
with his own small square of land,
                              some overseer's voice
echoing in abandoned mills where children
once worked sixteen-hour days, their ghosts
wavering, white-lipped, sharp-ribbed,
until I climbed over and, hands cut, found a road
          vaguely matching the map's traces.

Enough haunting for one day. But curiosity
wouldn't give in, and in Buxton's Micrarium
I went deeper, to countries more hidden:
          massively magnified
onto a screen, a segment of an orange
like an owl's head, a horse chestnut like a laughing frog,
the foot of a twenty-week-old human fetus—
the map of some lake district.

*

Keswick (Cumbria)

Get Ready to Ride the Lion to Zion
throbbed on the Walkman as the bus slipped
downhill, and in neat unison another Zion
flashed, a blue lake to answer my dry eye
thirsty from stone walls and stone-sided dales.
Against my lids, Coleridge and De Quincey—
dreadlocked, dressed in rainbow-striped smocks—
swayed to Jamaican rhythms soul-shakin', lion-proud.

Ankle blisters and hand cuts healing,
yesterday I read in the B & B. Did you know,
no early histories have anyone climbing mountains
for the view's sake?—
        a grounding shock
that ascents are no more instinct than speaking
a certain language or eating a certain food,
        gingerbread or beef.

Still, foot wrapped in a stretch bandage long enough
to circle a gorse bush, today I trekked ten miles
        through Borrowdale and up Skiddaw.
Two axes cut the morning calm:
    Air Force jets streaking so low I ducked
and an old purple-socked Cumbrian cursing
Sikhs and Arabs "swarming through the country"
(so much for Solitaries or Wanderers made wise
from years climbing the fells). In his words
our language was barbed wire or a sluggish river
dividing field from field. I hardly spoke.

*

Top Withens, Haworth (West Yorkshire)

Up through the cloud-smothered moor
 wind beat the bushes and few stunted trees,
  made demented music in my rubber raincoat!

What a spot to pull out paper and pen,
 within a gutted house: toppled
  roofless walls, stone blocks, twisted grass,

a far cry from Heathcliff Coffee House,
 Three Sisters Hotel and Bronte Footwear.
  All the way up to this wuthering height,

not one stranger or bird
 as if the wind had whirled them all away—
  the raincoat snapping, billowing,

thrashing my back. The cold cutting,
 I can't say much more except
  our love's never felt like an uphill

fight, wind-flogged, star-crossed, hound-haunted—
 we close and peaceful are as mighty
  as *we* torn with torment and fever,

passion fluent and moistened and bedewed
 as grand as passion split
  between two homes, nerve-wracked!—

your foolish unHeathcliffish hiker.

*

Llanberis (Wales)

I climbed to the top of Snowdon today.
Half a million others will make it this year.
We're owl indy-vee-jewels? I've climbed Snowdon
but I've never seen it,
        never held its full shape in my eyes
while mist, rain, and snow have hidden it
from the streets of town for three days—
a mountain taken on faith, sky-absorbed.

In the heights the numbers of sheep dwindled
to zero, the air went piebald with snow. Alone
I felt squeezed in the fist of the elements
before the first Celtic exiles climbed, but the joke
was on me, Old Blister-Foot: on the summit
dozens mobbed with rucksacks and anoraks, half blind
in the spinning snow. What dumb idlers
we would've been to miller, priest, and shepherd
when mountains were havens for thieves
where the many-named devil boiled up storms.

Zigzagging back over trailless white
found myself on a blade-edge ridge, scree
and a sheer drop nearby on both sides,
rocks growing slicker—one slip fatal,
the air emptied of all mirrors. Wished
I were a mountain goat, a hawk, or myself
by a fire. I became i. Might've simply crouched
and embraced a stone for hours, crooning.

A feather of love fell through my senses—
        lost, I turned, rushed, hunched over
        keeping close to broken earth
until a familiar wide plateau stretched,
a stranger signalled the way I'd missed—

freed from snow, down past the first tree,
the first sheep. At last I was headed
homeward, to the airport where all else
fades once I see you. A stick-legged lamb ran
so fast it fell, fell again,
                    almost somersaulted,
furiously eager, torn four ways at once.

Melted snow turned parts of the map to pulp.
Next trip's with you, or no trip. Up there
with the throat-tightening risk, the thin ridge,
the dream of holding on to a rock and never
letting go—
                you sent that feather that far.

Thanksgiving in an Old House

for Karen

This holiday is an empty stone well
encircled with rotted leaves. Fill it,
let subterranean streams rush in—

for the fuchsia's red blossoms on the sill,
Indonesian salad, ginger root dissolving
in our mouths, the nearby siren crying
how someone would save a life or a house,

for single-celled noctiluca
massed and phosphorescent
on the ocean surface at night,

for this house of pine planks and old brick
newly ours, handhewn ceiling beams still
telling stories of the axemen's swing,

for jumps to the unaxable,
the unphotographable, the unframable,

for her, the blessing I never expected,
her voice and hands, all of her, all,

for everything mating our laughter,
optimists "those who say bad weather's
better than no weather at all,"
for wind
rubbing our faces when we climb
to a ferry's top deck, in a gannet's wake,

for the lungs to lament, for all who curse
a ferry's death and a monster bridge's birth,

for rain on the roof (as long as it won't
leak any more),
for *as long as, even if,*
yet, still, dashes and parentheses,

for the Anarchist who gardens,
the daycare centre in the seniors' manor,
the keychain skeleton hanging in the hall
and every other unlikely pairing,

for ingenuity, ingenuity!–
Hearne and his men in 1770 cutting up their tents
for shoes; Muir in Alaska, 1880, scraping shavings
from his sled, tiny kindling to brew tea,

for all who read closely, all who read
carelessly (whoever slap-painted the sign
METHODIST CEMETERY BAKE SALE)--
for *brouhaha, dogsbody, rigmarole, bosky,*

for the bedrock that no done day can be undone,
for regret, so the next game flies fiercer,
for sleep without nightmares, that healer,
that fall into blankness from which we rise
into each other's arms like startled swimmers,

for the mouth and fingers to give thanks
and the Zimbabwean thumb-piano, its mighty pluck
ringing to all the corners of the room...